LIGHTWEIG
BIKES

GH00370448

LIGHTWEIGHT BIKES

Mick Woollett

SURREY COUNTY COUNCIL
WITHDRAWN FROM STOCK AND OFFERED FOR SALE ALL FAULTS BY SURREY COUNTY LIBRARY

SURREY COUNTY LIBRARY

B. T. Batsford Ltd, London

MOT/629.22

81-616621

Page 1
Belgian Harry Everts, 125 cc World
Champion 1980, flies high on his
water-cooled factory Suzuki.

Page 2
Honda's latest 250 cc sports roadster,
the CB250RS, is powered by a
single-cylinder four-stroke engine
and is considerably cheaper than
their best-selling 250 cc twin.

©Mick Woollett 1981
First published 1981

All rights reserved. No part of this
publication may be reproduced, in any
form or by any means, without
permission from the publishers

Typeset in 10/11pt Stymie Light by
Typewise Limited, Wembley
and printed in Hong Kong
for the publishers
B. T. Batsford Ltd
4 Fitzhardinge Street
London W1H 0AH

ISBN 0 7134 3913 0

CONTENTS

Pioneer Lightweights	7
Dazzling Choice	14
The Next Step	22
Ultimate Lightweights	30
Schoolboy Boom	38
Off-Road Upsurge	46
Power-Packed Lightweights	55
Index	64

Acknowledgment

The author and publishers would like to thank Jan Heese and *Motor Cycle Weekly* for permission to reproduce their photographs.

The advantages of a lightweight are
well illustrated in this lovely shot
taken near Mombasa in 1925. The
bike is a Francis Barnett.

PIONEER LIGHTWEIGHTS

It should be, you would think, the easiest thing in the world to define a lightweight motor cycle. After all, what else could the term mean than a motor cycle which is lighter than some given figures; 224 lb, let's say for the sake of argument. Ah, but life is never that simple. Over the years we have come to accept as a 'lightweight' any bike with an engine capacity of 250cc or less, sweeping aside the undoubted fact that a simple, single-cylinder 500cc model might well turn out to weigh less than some four-cylinder mini-superbike.

There is a historical precedent for accepting a 250cc upper limit, because when the Isle of Man TT Races first added to the programme an independent race for 250cc machines, it was given the plain and unvarnished title of the Lightweight TT. That was in 1922, and just for the record the winner was Geoff Davison, on a little two-stroke – an important point, to which we will return shortly.

A couple of years later, the organizers of the TT decided to introduce a race for still smaller machines, but since they were already using the 'Lightweight' title for the 250cc event, they had to think of some other name for the new 175cc class. The name they thought up was the Ultra-Lightweight TT, and *that* was to cause considerable confusion when, a couple of decades later, a 125cc racing class was instituted. And as for 50cc racers…!

But let's leave it at that for the moment and go back to the pioneer days of motor cycling. Around the 1900s, a motor cycle was usually just a pedal cycle with an engine clamped ahead of the front down tube and driving, by direct belt, to a pulley rim clipped to the spokes of the rear wheel. The cycle's pedalling mechanism was retained, because power outputs were generally low and what was described as 'light pedal assistance' had to be applied, to help the machine to climb anything that varied however imperceptibly from the dead level.

However there were engines and engines. The London-based Matchless company hung a whacking great MMC (Motor Manufacturing Company) engine on the front down tube of their pedal cycle frame, and went racing on the cycle board-tracks of Canning Town and Herne Hill. So did Excelsior, of Coventry.

Real lightweights had to wait until the arrival of a much smaller, and more efficient, clip-on engine manufactured in Belgium. This was the 1¾hp Minerva of 1901-2 which, initially, employed a mechanically operated exhaust valve in conjunction with an overhead 'automatic' inlet valve (AIV). The AIV arragement, quite common at the time, embraced a valve fitted with a very light spring, sucked open to admit fresh gas by the action of the descending piston.

Naturally, AIV could only work satisfactorily at lowish rpm, but by 1902 Minerva had evolved a much more sophisticated 1¾hp engine, equipped with mechanically operated side-by-side valves. It was this little clip-on side-valve, the first unit of its size to possess any degree of reliability, that took Triumph, among others, into the world of motor cycle manufacture.

By the way, don't be alarmed by that '1¾hp' figure. It was an arbitrary quotation and bore little relation to the actual power output of the engine. Taxation, once it got into its stride, was by a motor cycle's overall weight, not its engine size, and riders and makers got into the habit of grouping bikes into horse-power pigeonholes. The 1¾hp Minerva was actually 211cc, but in fact within a few years a 211cc engine would be more usually counted as a 'Two-and-a-quarter'.

Similarly, a 'Two-and-a-half' could be anything from 244 to 296cc, at the maker's discretion, but a 'Two-and-three-quarters' was almost invariably a 350cc, while a 'Three-and-a-half' was a 500cc.

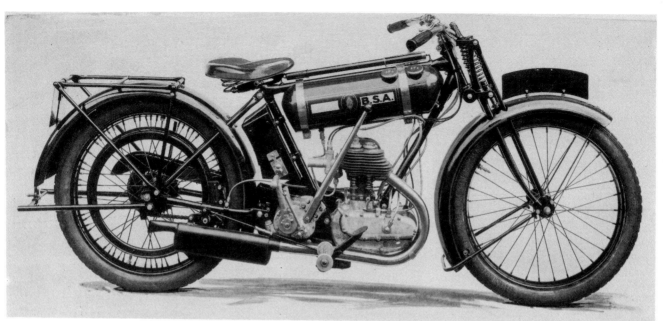

Famous lightweight of the twenties, the 250 cc BSA Round Tank.

One of the better lightweights of the thirties, the sporty little 250 cc New Imperial of 1936.

In the hands of determined riders, those early lightweights could be coaxed to perform quite remarkable feats, and it is on record that as early as 1900 a 1¼ hp Werner was ridden by Hubert Egerton (later a partner in a Norwich motor business) from John O'Groats to Land's End in four days and nine hours. Others took it in turn to chip away at that figure, until by 1910 the so-called Lightweight End-to-End Record stood to the credit of carburettor pioneer Harold Cox (1¼ hp Singer) with a time of 57 hours 26 minutes.

That July, however, Eli Clark of Bristol was to set the cat among the pigeons. His time for the John O'Groats to Land's End run was 39 hours 40 minutes, which was fair enough except that the machine was a 348 cc Douglas flat twin side-valve. In *The Motor Cycle* for 4 August 1910, the Douglas factory (makers of 'The Best of

Forerunner of today's mopeds - 98 cc James of 1938 which cost £18.90!

K18

ANNOUNCING

The PRINCE LIGHTWEIGHT

ROYAL ENFIELD

125. ENGINE CAPACITY
150 MILES PER GALLON
60 M.P.H...

FIRST & FOREMOST IN
RELIABILITY
& ECONOMY
£3126

£32.10.6. Deposit & payments over 2 YEARS.

George Formby
and his wife Beryl
with the 125 cc
Royal Enfield of
1951.

Lightweight at war.
This 125 cc James
was developed for
troops to use in
France after the
invasion in June
1944.

Royal Enfield Crusader Sports of
1960. A neat machine but it was not
a great success.

All Lightweights) advertised Clark's effort as 'The Lightweight End-to-End Record secured by a Douglas Lightweight Motor Bicycle.'

That really did set the correspondence columns alight. How could a three-fifty be a lightweight?, asked readers. Because it was a light weight, answered Douglas. But the Singer folk weren't having that, and they responded with an advertisement claiming that 'The Singer Moto-Velo established and still holds the Lightweight record – *for motor cycles of less than 1½hp.*'

'Deeds, not words, make a reputation,' retorted Douglas, pointing out that their 350cc twin was 'Guaranteed under 112lb in weight,' and certainly it would seem that clubs were recognizing machines of up to 120lb as being lightweights for competition purposes.

However, let's get back to the under-250cc class, because around 1911 or so, the utility two-stroke began to make headway. Leading the pack was Levis, the very name of which was the Latin word for 'light,' while to emphasize the point, the Levis headstock badge depicted a pair of crossed feathers. Originator of the Levis was Bob Newey, who worked at the Norton factory, and his first two-stroke lightweight was in fact built in the Norton plant. Bob showed his baby to James Lansdowne Norton in person but, alas, Mr Norton didn't want to know. 'Take it away,' said he. 'We make *men's* machines here!'

And that might have been the end of the matter, but it happened that Bob Newey was courting a Miss Daisy Butterfield, whose two brothers felt that the neat little bike had something. 'Let's make it ourselves,' they decided, and that's just what they did. The 211cc Levis rapidly gained a big following, and the same engine design was adopted also by New Hudson a little while later. And following the First World War it was the Levis, made under licence by Zundapp, which helped to get the German motor cycle industry back into production. But that was still to come.

More companies began to get involved in the lightweight two-stroke scene. There was Connaught, the first to hit on the idea of lubricating a two-stroke by tipping a measured quantity of oil into the fuel. And there was Triumph, with the very pretty little 225cc Junior, without kickstart or clutch but incorporating a two-speed gear.

Indeed, it was very rare for any light two-stroke to feature a clutch or kickstart. One just sat astride, pulled the compression release-valve trigger and paddled off with both feet, when the motor would usually fire as soon as the release valve trigger was dropped. At traffic halts – and they were few, because traffic was almost non-existent – it was simply a matter of stopping the engine, and paddling off again when the road was free.

As often as not, the cheapest lightweights had no gearbox at all, and transmission was by a rubber-and-canvas vee-belt (perfectly capable of transmitting the miniscule power output of a wee two-stroke) direct from engine pulley to the rear wheel.

It was in the years immediately after the First World War that the lightweight really came into its own, and that was largely because specialist engine makers such as Villiers, T. D. Cross, Liberty or Union could offer quantity-built power units to small assemblers at attractive prices. A couple of men could hire a small workshop, order a small batch of engines from Villiers; some gearboxes from Albion, Burman, or Sturmey-Archer; front forks from Brampton, Gosport, Maplestone, or Druid; sets of frame lugs from Haden or Chater-Lea; and presto! A new make was born!

They were two-strokes, mainly for simplicity, but also because steel technology was still in its infancy. Overhead-valve engines were mistrusted, because valve heads

tended to break off and drop into the combustion chamber with expensive results – not so much a problem with bigger engines, when the valves could be made suitably beefy, but valves for lightweights had of necessity to be tiny.

Nevertheless, technology was advancing, and perhaps the favourite lightweight of the mid-twenties, the 249cc BSA Round Tank, employed a side-valve engine. By 1932, the steels were available to permit valves to be made for even 150cc overhead-valve engines, and some very smart little o.h.v. lightweights in this capacity emerged from New Imperial (possibly the best of the breed), Royal Enfield, Triumph, BSA and Excelsior.

But there, maybe, we can end this historical survey. Lightweights of the thirties had come a long, long way from the pioneer clip-ons. Now, they were all-chain drive, with specifications that included (in some cases) four-speed gearboxes with foot-change. Gas lighting had given way to electric. So what else was there to be developed?

Well…telescopic forks, and rear springing, and dualseats, and alternators, and overhead-camshafts. But really, these were incidentals, and a 1939 lightweight wouldn't look too out-of-place on the roads, even today. Have a look at some at a vintage gathering, and you'll reckon so, too.

DAZZLING CHOICE

Today's prospective buyer of a lightweight motor cycle is faced with a dazzling display of machines ranging from the simplest of mopeds to the ultra sophisticated 100 mph models that top the 250cc ranges of the world's leading manufacturers.

But let us look first at the types of machines with which most riders start their two-wheeled careers – at the restricted mopeds which riders of sixteen are allowed to ride in the UK and in the majority of European countries and at the new wave of small-wheeled semi scooters which are proving such an attractive buy for the commuter, student and housewife.

Although it was the European factories which initially developed the 50cc moped and who still make the basic, powered bicycle type of vehicle in vast numbers (the French Motobecane and Mobylette factories remain the world's largest makers of down-to-earth 50cc mopeds with a combined output of around 500,000 units a year) it is the Japanese who have led the way with the more sophisticated and glamorous machines which have revolutionized the British market.

It was the Japanese who developed the 50cc sports mopeds of the mid-seventies – miniature motor cycles with five-speed foot-change gear-

boxes, good brakes, comfortable saddles, eye-catching styling and a performance equal to that of the 125cc machines of the sixties.

The fierce competition for the booming market which the Japanese had created led to the development of machines that were, according to the authorities, simply too fast for novice riders. So in the late seventies a law was passed in the UK to restrict sixteen year olds to a low performance category of machines – known in the trade as a restricted moped or a 'sixteener.'

At first the manufacturers were caught at a disadvantage. But since then they have responded by developing bikes with improved looks and comfort and big bike features designed to compensate for the 30 mph top speed to which these machines are limited.

They have also developed new

Impressive newcomer from Kawasaki – the AE50 introduced for 1981 and aimed at the booming trail bike market.

American style! Plugging every gap in the market, Suzuki offer this customized model, the OR50.

models disigned to appeal to special sections of the market. For example, for 1981 Suzuki offer five different 50 cc models for the sixteener to ride: a smartly styled roadster, an American styled chopper version, a dual-purpose road and trail model and two scooterettes – one with extensive mudguarding and cowling, the other a cut-price, no nonsense, naked version.

Yamaha, who led the way in the development of the original sports moped, have a similar range headed by two normal roadster models – the up-market RD50M with all the features one expects to find on larger motor cycles and the cheaper FS1DX, developed from the original moped.

Honda, who lost out heavily in the sports moped market because they stuck to four-stroke designs which were more complicated and less powerful (and therefore slower) than the rival two-strokes, have responded with a completely new range of purpose built two-strokes which have put them back into the market with a bang.

The European manufacturers, who once led the way, are hard pressed to compete with the Japanese. However, the East European makers, hungry for hard Western currency, offer good value for money to riders looking for down-to-earth transport without the frills – MZ of East Germany and Jawa of Czechoslovakia both offering models that are worth considering.

All these 50 cc restricted mopeds are now powered by basically similar engines – by simple to make and easy to maintain two-strokes. To cut out the sometimes messy business of mixing the oil with the petrol, a weakness of older two-stroke designs, the Japanese have developed systems with a pump and a separate oil tank.

Because two-stroke fuel pumps and fore-court mixers are still common in Europe, the European makers have stuck to the older system. It is a small point to be considered when choosing a machine, but what about the other points?

A major factor, and one which first-time buyers tend to ignore, is the availability of service. Is the dealer you are buying from reasonably near your home and has he the facilities to repair your machine if you are unfortunate

Typical of the trend into trail bikes, the 50 cc Yamaha DT50M.

Czech challenger for the sports moped market, the Jawa Mustang.

enough to run into trouble?

Obviously the novice is more likely to need his dealer's help than the experienced rider who, over the years, will have learned to rectify minor faults and will have bought the necessary tools to tackle small jobs.

So my advice to the first-time buyer who does not know a great deal about bikes is: go along to your local dealer. Have a quiet look round and assess the place. Is it merely a shop where they want to sell you a motor cycle and get you out of the way or do they offer a genuine service backed up by workshop facilities?

If it is all flashy shop-front and smartly dressed salesmen, beware. The motor cycle business has boomed in recent years and some have jumped on the band-wagon. But the majority of dealers are well established and a chat with a couple of local motor cyclists should help you to decide which dealer you should go to for your new, or secondhand, machine.

Of course, there can be complications if you have set your heart on a particular model which your carefully selected dealer does not stock. If that turns out to be the case I should have a

Popular roadster – the Yamaha RD50M, latest in a long line of mopeds from this famous Japanese factory.

Kawasaki entered the 50 cc field for the first time in 1981 with this sporty looking **AR50** model.

good look at what he can offer. The top models are now so similar in price, performance and reliability that it is probably more important for the first-time buyer to have a good, sympathetic dealer than a particular make of machine.

The same applies to discounts. It is very tempting to buy from a dealer a hundred miles away if he offers a bike

The little bike that started a trend –
the 50 cc Honda Express.

at £20 or £30 less than your local man. But by the time you have fetched it and maybe had problems with the first service (normally provided free by the dealer who sells the machine) you will begin to wonder if it was worth it.

These guidelines are particularly valid if you are considering buying one of the new wave of runabout 50 cc machines – the small wheeled bikes that are a cross between the Lambretta/Vespa type scooters that were so popular in the sixties, and the normal moped.

Honda first tapped what is proving to be a tremendous market when they introduced their stark little NC50K1 Express. Backed by an extensive advertising campaign, which included a television commercial featuring Twiggie, and selling for below the cost of most mopeds, the Express's popularity took off like a rocket and sold 19,000 in the UK in 1979 – 70 per cent bought by women for commuter-shopping transport.

Why was the concept so successful? Probably because the Express looked as easy to manage as a bicycle and cost only twice as much – and of course it was launched, by luck, at a time

Suzuki's answer to the Honda
Express, the neat little FZ50 with two-
speed automatic transmission.

when petrol prices literally doubled within the space of two years.

The Express is powered by a single-cylinder two-stroke engine. There are no gears or clutch. You simply start the engine with the unique wind-up starter system (you depress a pedal a number of times to wind up the mechanism, then trigger it off to start), open the throttle and away you go – the automatic clutch taking up the drive as the throttle is opened and the engine revs rise.

To keep the drive simple and smooth a belt is used instead of a chain. After Honda's initial success, Suzuki followed suit with a slightly bigger and more sophisticated model, the FZ50. This is more expensive but you get a lot for the extra cash and the Suzuki has the great advantage of a two-speed automatic transmission which makes it more suitable for longer journeys.

This model proved a great success too. Suzuki sold their first batch so fast that dealers were out of stock in days, and then had to wait months while the importers put in new orders for the machines to be shipped from Japan.

Simple, automatic mopeds which cover up to 185 miles per gallon (like the Vespa Ciao) make economic sense for shopping and commuting short distances.

Full-sized Vespa scooter, yet only powered by a 50 cc engine – the Vespa 50 Special.

One of the great problems facing the Japanese makers is that they have to forecast what will sell in Europe over 12 months in advance to allow for making the machines and then shipping them!

Jumping on the bandwagon, and taking the styling another step closer to the original scooter idea, Yamaha introduced their challenger in this new field, the Passola, in 1979. This machine was also a success, helping to push Yamaha sales that year to a new high record.

Rounding out the 50 cc models available are a number of mopeds, ranging from the traditional upright style, Austrian-built Puch Maxi, to the full Vespa scooter, the capacity reduced from the traditional 125 cc so that it can be ridden in its native Italy in a special low-powered class and sold in the UK as a restricted moped.

A dazzling choice indeed – totalling close to 50 models. But, for value for your hard-earned money, the big companies, and particularly the Japanese, are hard to beat.

THE NEXT STEP

Are you seventeen or over and considering a bigger machine? Then the next step is to look at the 100 cc and 125 cc bikes - the smaller capacity now very popular because of the considerably lower insurance premiums applicable to bikes under 100 cc.

Again, the choice is tremendous and it is complicated by the fact that as the engines get bigger, so the specifications broaden. Four-strokes and twin-cylinder engines begin to enter the story, although the great majority in the class are simple, single-cylinder two-strokes.

Years ago manufacturers used to offer just one or two models in this class. Currently Suzuki list ten, Honda nine, Yamaha seven and Kawasaki six - so the prospective buyer has a choice of 32 machines in this class from the Japanese manufacturers alone!

Broadly they fall into four classes - step-through super mopeds of which the Honda 90 is the best known; no-frills cummuter bikes like the Suzuki Graduate; sports roadsters of which the twin-cylinder Yamaha RD125 is a typical example, and dual purpose road/trail bikes like Kawasaki's very sporty-looking KE125-A7.

In addition to the bikes offered by the Japanese big four there are also a number of European machines in this class worth considering. For example, the rider with a limited budget looking for a sturdy, reliable cummuter should study the 125 cc MZ Alpine. This is

Cut-price commuter – Suzuki's 100 cc Graduate with rotary valve two-stroke engine.

Best of the Eastern bloc machines is the MZ. This is the 150 cc model marketed in the UK as the Eagle.

built in the old pre-war DKW works in East Germany and normally retails at around £100 less than the cheapest Japanese machine in the class.

At the other end of the scale the Italian-built bikes tend to be every bit as well equipped as the Japanese but more expensive. Prime examples are the Gilera 125TGI, a neatly styled, sporty single-cylinder two-stroke, and the Benelli 125 Sport, a rorty little twin-cylinder two-stroke that will top 80 mph – but costs a lot of money!

There is even a British-built bike in the class, the 125cc BSA Tracker. However, it is only fair to point out that the engine is basically a Yamaha; that many of the cycle parts are Italian and that it costs more than the Yamaha equivalent – you have to pay to be slightly patriotic!

Faced with such a bewildering choice, what should you do? First of all you have got to sit down and decide exactly what you want the bike for – what will it be doing most of the time that it is in use? It may be simply to take you shopping or for short trips to your place of work, in which case you are faced with a choice between the step-through scooterettes offered by Honda, Yamaha and Suzuki, and the simple, commuter motor cycles listed by all manufacturers in the class.

For the first-time buyer, and for the slightly older buyer not worried about his image, the step-through scooterette style machine with its comprehensive built-in weather protection must be a favourite; especially as these machines normally retail for considerably less than the equivalent commuter motor cycle.

As an example, the superb little Honda C90ZZ, powered by a 90 cc overhead camshaft engine and with a three-speed gearbox, sells for £40 less than the new Honda H100A, a single-cylinder, two-stroke general purpose motor cycle.

But of course there are a lot of people, especially youngsters, who simply will not consider the scooterette type of bike with its district nurse, elderly farm worker image. If they are going to buy a motor cycle it must look like a motor cycle.

Fair enough. It is no good having a bike if you are not happy riding it – the choice now is between simple commuter, sports roadster or trail bike. Because of the specification, the simple commuters are obviously cheaper and it is a question of looking around to see what you like the look of and where you can get it.

Here too you are likely to be dependent on your local dealer for after-sales service and it is well worth talking to him to get his advice. If he knows you are a local person he is most unlikely to try to sell you a bike which is liable to give trouble. The dealer's aim

Benelli's 125 Sports – a high performance two-stroke twin with race-bred roadholding.

capacity machine – and a top speed of around 75 mph means that it can match a lot of bigger bikes on the road.

But remember that performance and sporty looks usually mean that the bike is harder to ride and may be less comfortable. This is because when you go for top power in an engine you sacrifice flexibility and the ability to

is to make a profit – and to end up with satisfied customers.

To achieve both he has to sell bikes which give the minimum amount of trouble. Customers who keep coming back with machines that have to be repaired under guarantee force the dealer to spend quickly any money he may have made on the original sale. Therefore it is in his interests to sell you a machine that will give you no trouble.

Younger riders will be attracted by the sports roadsters. Unchallenged king of this class is the Yamaha RD125, a scaled-down version of its famous big brother, the RD250, and a fine motor cycle from any point of view. Powered by a twin-cylinder two-stroke engine with five-speed gearbox, disc front brake, big tank and full-sized dual seat, it can easily be mistaken for a larger

Ideal commuter machine – the Honda H100-A with single-cylinder two-stroke engine.

24

Little racer! The Italian Beta 125 cc
two-stroke is a joy to ride on the open
road.

Handsome machine – the Suzuki
TS125 trail bike.

Popular model from Suzuki – the
GP100 Sport.

pull at low revs (torque), and when it comes to rider comfort the sports bike usually has a semi-racer riding position with rear-set footrests and low-slung handlebars.

The fourth type of machine, the trail bike, is the fastest growing category and has a lot to recommend it. Not only does the slightly sit-up-and-beg riding position mean that it is comfortable to ride and easy to manoeuvre in tightly packed city traffic but it gives the rider the chance to explore the countryside by riding on what are known as green lanes – unpaved lanes now mainly used as footpaths or bridleways but which are still legally open to traffic.

The trail bike was originally developed, mainly in the United States, for the rider who wanted a dual-purpose machine – for road and off-road use. As with all compromises, trail bikes have their faults. They are not suitable for long-range, fast road work – neither are they ideal for really rough cross country trips.

But as a general-purpose motor cycle for rides to work and general pottering around they are hard to beat – especially now that the major manufacturers have realized the potential of the market and offer really

well equipped, strikingly handsome machines at reasonable prices.

Four-stroke or two-stroke? My own preference is still for the two-stroke. 125 cc engines are on the small side for the advantages of the four-stroke to outweigh the simplicity and robustness of the modern stroker.

One point worth remembering when you are buying is that you will eventually want to sell the machine. And it is the well known model from the well known maker that holds its value best and is easiest to sell or to trade in against a new model, when the time comes to get a new bike.

Superbly styled – the Gilera 125 TGI from Italy.

ULTIMATE LIGHTWEIGHTS

Right from the earliest days of motor cycling the 250cc has been a favourite class and although the image is a little blurred now by the proliferation of 175cc and 200cc models, it still marks the boundary line for the lightweight division.

The choice is tremendous. At the start of 1981 there were over 60 different models from over ten manufacturers that fell into the over-125cc and under-250cc capacity bracket.

Benelli and Yamaha vie at the top of the market with bikes for the person who must have the ultimate in the class. Benelli offer a four-cylinder, overhead camshaft, five-speed dazzler that is tremendous fun to ride and capable of very close to 100 mph – a road-going version of their famous racer of the sixties.

Yamaha have countered with the bike that was the talk of 1980 – the RD250LC. This too is a roadster developed from a successful racing machine, the TZ250 Yamaha which in standard form is simply the best 250cc racer which money can buy.

Top lightweight road-burner – the race-bred Yamaha RD250LC with water-cooled, twin-cylinder engine and a genuine top speed of over 100 mph.

The RD250LC is powered by a water-cooled, twin-cylinder, two-stroke engine with six-speed gearbox and enough power to make it the first genuine 100 mph lightweight ever offered for sale to the public. And although it is expensive, at just over £1000, it is very much cheaper than the Benelli.

However, please do not buy either if you are a first-time buyer – these are high-performance sports machines and are definitely for experienced owners only. Novices on bikes like these very quickly get into accident statistics and give motor cycling a bad name.

To complement the RD250LC, Yamaha also offer an air-cooled two-stroke twin, the RD250, a 200cc version (RD200) and two neatly styled four-stroke twins, the conventional XS250 and the Americanized chopper-style XS250S which for a few pounds more gives you that easy-rider, laid-back look which is ideal for high street cruising, but not very practical for fast road work.

Tremendous fun. The four-cylinder, overhead camshaft Benelli 254 (250cc, four cylinder) that revs to 11,000, has a top speed of over 90 mph and handles like a racer.

Yamaha's XS250 is a down-to-earth roadster equally at home in town or country.

Yankee style! The Suzuki GS250TT, introduced for 1981, has American laid back styling allied to a high performance sports engine.

In fact the success of these customized machines with their high handlebars, stepped 'king and queen' dual seats and small, rounded fuel tanks took the Japanese companies by surprise when they introduced them on the American market in the late seventies.

For they thought that the demand for these machines would be far less than for the conventional, cheaper models from which they are derived. But in some cases they actually outsold the originals, a situation which caused a hiccup in supplies, with custom models selling out and the standards becoming a glut on the market.

In addition to their two-stroke and four-stroke twins, Yamaha also offer two single-cylinder four-strokes, the roadster SR250S and the trail bike XT250, both powered by basically the same single overhead camshaft, five-speed engine.

Like Yamaha, Suzuki, who pioneered the super-sports two-stroke twin concept with their Super Six model of the sixties, offer a full range of four-strokes and two-strokes, twins and singles.

Their most impressive model is the GSX250, a superbly styled, double overhead camshaft, four valves per cylinder, four-stroke that produces 27hp at 10,000 revs – an excellent sports model for the rider who wants to travel far and fast with style.

The latest version of the old Super Six in the high performance two-stroke stakes is the GT250X7, slightly faster than the GSX250 but more tiring to ride and not such a good all-rounder as the four-stroke.

Recently Suzuki have favoured the trail bike theme, and these aggressively styled machines with their brilliant paint jobs in contrasting colours are certain to catch the eye in the showroom – including the TS250 and the TS185 that fall in this capacity group. Both are single-cylinder two-strokes with five-speed gearboxes.

Little thoroughbred. The Yamaha SR250 is an overhead camshaft single-cylinder lightweight that is a joy to ride.

Top model in Suzuki's 250cc range is
the sports roadster GSX250.

Mid-range trail model, the Suzuki
TS185ER is ideal for short
commuting trips and for exploring
the bye-ways at weekends.

Kawasaki, smallest of the Japanese manufacturers, are unique in offering a three-cylinder two-stroke, the popular KH250 which offers sparkling performance coupled to appalling fuel consumption – driven hard it covers only 40 miles per gallon!

It was originally developed for the American market at a time when fuel economy did not matter. Overtaken by events, it is still in production as a super-sports lightweight, but Kawa-saki have developed a whole range of other lightweights.

Top roadster is the neat little Z250A sold in the UK as the Scorpion. This is powered by an overhead camshaft, twin-cylinder engine which produces 27 hp. If money is tight an economy model, the Z250B, powered by the same engine, is also available.

For the commuter, Kawasaki offer two very nice single-cylinder four-strokes – the Z200 and the Z250C. Two trail bikes round out the range which includes something for just about everybody!

Market leaders in sales of the class are Honda, the Japanese giant who now make over two million motor cycles a year. They suffered in the mid-seventies because they lacked glamour machines to match the new Yamahas and Suzukis.

Then they introduced the Dream, a 250 cc overhead camshaft twin, and

Best seller. The Honda CB250N
Super Dream topped the 250 cc sales
chart in 1980.

sales took off. Part of this machine's
appeal is that it is, by 250 cc standards,
a big bike. It offers a lot for the money
and the new European version is a
very good looking bike—definitely the
top selling 250 cc on the British
market.

The trouble is that it is, like its direct
rivals, rather expensive. So in 1980
Honda introduced a single-cylinder,
overhead camshaft, four-stroke, the
CB250RS which sells for about £140
less than the twin but has a very similar
performance.

Slow to enter the trail market, Honda
have recently really moved into this
field and now offer superb-looking
180 cc and 250 cc singles.

At the budget end of the market, MZ
of East Germany and Jawa of
Czechoslovakia both offer two-strokes
at well below the prices of their
Japanese competitors. As with the
smaller machines, the MZ is the better
buy. In fact as a commuter-cum-tourer
it is hard to beat.

You may also be tempted by low-
priced Russian machines which
appear on the market from time to time.
In the past these have proved to be
poorly made and unreliable – and as a
consequence virtually impossible to

sell once you realize you have made a mistake!

For the person who wants something different there are a handful of models produced by the smaller European makers. Most notable among them is the delightful little Morini four-stroke vee-twin from Italy; the single-cylinder two-stroke Cagiva from the same country and the super-sports KS175 Zundapp, a little thoroughbred from the famous old West German factory.

But to be different costs money. These models are priced well above the opposition, and service and availability of spares should be carefully considered when you are buying.

Value for money. The 250cc MZ is well designed and is budget priced at well under the Japanese opposition.

SCHOOLBOY BOOM

The boom in the sales of schoolboy bikes is one of the major plus factors that has made the motor cycle business such a lucrative one in recent years. Ten years ago the trade in special machines for children was just a trickle. Now it is a multi-million pound business with bikes ranging from the simplest 50 cc automatic to ultra high-powered 125 cc moto cross machines

Latest small bike from Yamaha is the PW50. Designed as a fun bike, it features shaft drive.

which would not be disgraced in a fully fledged grand prix.

At the bottom of the range are the simple 'garden bikes.' These are designed to be ridden by children from just four years old up to eleven or so and are normally powered by single-cylinder two-stroke engines with automatic transmission – the rider simply starts the engine, opens the throttle and away he/she goes without having to worry about clutch or gearchanging.

These little machines are robust, easy to ride and service and provide endless hours of fun for the proud owner and friends – and while quick enough to provide the necessary thrills, seldom have a top speed of over 25 mph.

They are of course not legal on the road and may not be ridden in a public place. But, for a large family garden with a rough patch of ground which will be torn up by the knobbly tyres, they are the ultimate toy for the mechanically or speed-minded child – which covers just about all of the species; for there cannot be many youngsters who would not love to ride a motor cycle given the chance.

The Italians were the first into the market with Technomoto, Ital-jet and Malaguti leading the way – all using engines developed by the Italian moped industry. The Austrian Puch factory spotted the potential and scooped a large slice of the British market when they introduced the Magnum, a superbly engineered and very rugged little bike which sold so well that the British importers were soon clamouring for more.

Interestingly Puch broadened their sales base by selling the Magnum through garden centres, lawn mower shops and so on – anywhere where people with large gardens were likely to go.

The next step for youngsters is to

Sheer bliss! The writer's son Guy gets to grips with the Puch Magnum, a 50cc automatic.

move onto the 50cc bikes with clutch and gearbox, real little scaled-down moto cross machines. Riding these a child really does learn how to control a powered vehicle and this knowledge is certain to be useful when he does eventually move onto the road whether it be on a bike or as a car driver.

He also learns how to handle a motor cycle on slippery and bumpy ground, knowledge that will give him confidence and the ability to get out of trouble when he does progress to the road. So these first schoolboy bikes are very definitely educational as well as tremendous fun.

Up to now the Japanese have left the small bike end of the market to the Continentals. They have concentrated on providing the schoolboy competition riders with sophisticated potential race winners, the best of which cost well over £1000!

Far and away the most popular branch of schoolboy motor cycle sport is scrambling (also known as moto cross and consisting of racing around a grass or mud surfaced track with left and right hand bends and with undulations, jumps and ditches).

There are normally four main classes in all forms of schoolboy motor cycle sport in which boys of certain ages are limited to engines of certain

sizes. In moto cross the youngest are known as Cadets, are aged six or seven and may ride only 50 cc machines. Next come the Juniors (age limit ten) who race bikes up to 80 cc. They progress to being Intermediates allowed to race 100 cc machines (or 150 cc if using a British engine) between the ages of eleven and thirteen.

After that they move into the Senior division which is sub-divided into two: class A for fourteen and fifteen year olds and class A1 for the Senior Experts aged fifteen and sixteen. Both sub-divisions race 125 cc foreign-built machines or 200 cc British.

Suzuki were the first into the field and their RM range of schoolboy scramblers still sets the standards by which others are judged. Serving the four classes, they come in capacity sizes 50 cc (RM50), 80 cc (RM80), 100 cc (RM100) and the 125 cc class for which a brand new water-cooled version, based on the factory's successful adult grand prix machine, was introduced for 1981.

Yamaha too have a full range and introduced new models at the Cologne Show late in 1980. They differ from the Suzukis in having the monoshock rear suspension system pioneered by Yamaha in moto cross and road racing. Instead of the orthodox pivoted rear fork with a suspension unit each side of the rear wheel, the Yamaha layout relies on a bigger, single unit mounted near horizontally under the fuel tank

Typical youngsters' fun bike, the 50 cc Malaguti Ronchino is suitable for children from six to ten.

Miniature moto crosser. Although the Italian-built Malaguti CR Arrow 50 cc machine looks like a big bike, it is in fact small enough to be raced by six and seven year olds.

The Suzuki RM80 looks just what it is – a potent 80cc moto cross machine.

and connected to a strongly braced double rear fork.

Yamaha claim that this system is stronger, gives greater suspension movement (very important on rough tracks) and because the suspension unit is bigger, it is less likely to overheat and change its characteristics as the race goes on.

The smaller models are powered by single-cylinder air-cooled, two-stroke engines with the now popular reed valve induction system, designed to give better power at low revs, driving via six-speed gearboxes.

However, like their rivals, Yamaha have been forced to water cool their latest 125cc model, the YZ125. This is because, as power outputs go up, so the engines have to dissipate more heat – and there comes a time when the familiar fins on the cylinder barrel and head just cannot cope.

The way out of the problem is to encase the barrel and head in a water-jacket and to add a radiator and pump to circulate the coolant. This system has allowed Yamaha to push the power of their 125cc moto crosser up to 30 hp at 10,500 revs; phenomenal for such a

For 1981 the Yamaha 125cc moto crosser is water-cooled with the radiator on top of the forks (the air-intake looks like a headlamp).

Top schoolboy moto cross machine, the Kawasaki KX125, looks the part.

tiny engine.

But moto cross is not only about power. Often a wide spread of usable power is preferable to a 'peaky' engine which gives power only at very high revs at the expense of torque throughout the rev range – and the ability to translate the power available into speed on the circuit is equally vital.

In their attempts to do this, Kawasaki claim that they have improved on the Yamaha monoshock layout by perfecting their Uni-Trak system. This too uses a single, large, suspension unit but unlike the Yamaha, it is mounted vertically near the gearbox, under the

seat. This is connected to the rear fork by a linkage incorporating a rocker arm – and it is this pivoted arm which is the secret of this system's success. For by altering it, the rear suspension characteristics can be changed easily and quickly when the machine is being set up for a particular circuit.

So far Kawasaki have stuck to air-cooling for their production models but they are experimenting with water-cooled factory bikes. Unlike Suzuki and Yamaha they have no 50 cc bike, preferring to concentrate on the bigger classes.

Busy trying to satisfy the world

demand for their road bikes, Honda are only now taking an interest in the schoolboy scene. Surprisingly they offer no garden bikes and the only scrambler is their 125 cc model, produced for adult moto crossers.

Other forms of schoolboy sport for which lightweight motor cycles are used are grass track racing and trials. Grass track racing is like speedway racing and is held on short, flat, oval tracks around 400 yards to the lap.

Because the circuits are smooth, short and have no jumps, the machines can be lighter and with far less suspension movement. Unlike

43

Young grass trackers listen attentively as an instructor gives them a word of advice before a race.

moto cross none of the big manufacturers make ready-to-race grass track machines. So competitors build their own, using frame kits from small, specialist manufacturers such as Alf Hagon of Leytonstone, fitting the engine of their choice, usually a moto cross unit.

Trials are a completely different form of motor cycle sport. Speed does not come into it. Instead competitors are pitted, one at a time, against hazards which they have to traverse without touching the ground with their feet. One touch of the foot is known as a dab and the penalty is one mark lost. More than a dab becomes footing and three marks lost. Ultimate sanction is the loss of five marks for a stop.

Obviously a light machine able to trickle along at less than walking pace, yet able to accelerate briskly and cleanly, is the best tool for the job. So far only Yamaha of the Japanese makers have moved into this field, with their 80cc Whitehawk model. This is actually built in England especially for the schoolboy market, and is proving a popular model, competing against a number of Italian machines originally produced for adults, imported and modified for schoolboys. Best known are the Fantic and Ital-jet models.

Schoolboy motor cycling has already produced one World Champion for Great Britain – Graham Noyce who started his career in scrambling as a teenager and went on to win the 500cc Moto Cross World Championship on a Honda in 1979. This lifted him into the £100,000 a year income bracket – schoolboy sport can lead to riches as well as fun and fame.

Graham Noyce, a schoolboy scrambler who made good, leads the pack on his Honda at the British Grand Prix.

OFF-ROAD UPSURGE

Off-road competition on lightweight motor cycles has boomed tremendously during the past decade – and looks set to go on growing despite the ever increasing problems of finding suitable terrain in a world that is becoming more and more crowded, and noise- and pollution-conscious.

Most popular of the off-road motor cycling pastimes is moto cross in which fields of up to 50 competitors per race battle it out over circuits that take in every sort of natural hazard from water to deep sand. The tracks are usually about a mile in length, bounded by ropes and stakes and with plenty of slow corners to keep speeds within reason.

The more hills the better and if no natural gradients can be found, enterprising organizers have been known to call in a bull-dozer to scoop out a few 'bomb holes' – in fact in the United States, where the sport has boomed tremendously in recent years, a whole new branch of moto cross has developed.

This is known as stadium cross and has been staged at some of the United States' most famous sports complexes – both outdoors and indoors! The idea is to bring moto cross to the American sporting public instead of trying to get the spectators to make the trip to some out-of-the-way circuit that is used only a few times a year and totally lacks the amenities and comforts which the American public demands: comfortable seats, hot-dog stands, cool drinks, good toilet facilities – and all much closer to home than a normal moto cross track.

How do you run a moto cross in a football stadium or inside an ice hockey arena? Simple, you bring in hundreds of tons of earth and build a track. And if you run out of space in such limited confines you take out some of the tiers of spectator seats and run the circuit up and down the terraces.

It costs a lot of money but it has been a success in the United States where crowds of up to 60,000 have packed famous sports venues, including the Los Angeles Coliseum and the Houston Astrodome for championship events. As a spokesman for the American Motorcyclist Association said: 'It may not be real moto cross, but whatever it is, it isn't bad!'

The sport started in England and Belgium in the twenties, got established in the thirties and boomed in the post-war forties and fifties, especially on the Continent. It quickly

Typical moto cross action as two Continental riders battle it out at the Dutch 125 cc Moto Cross Grand Prix.

spread east to the Soviet Union and then west to the United States until now, when there are certainly well over 100,000 active competitors.

Initially the machines used were 500 cc, mainly single-cylinder four-strokes. But as the lightweight motor cycle developed, so they began to creep into the sport, often surprising riders of bigger, heavier machines because they were less fatiguing to race and easier to manoeuvre around corners.

The lightweights progressed so fast that in 1957 a European Championship was launched. It was won by the West German Fritz Betzelbacher on a 250 cc Maico – a make still challenging for honours in the sport.

Britain had a brief reign of success when Dave Bickers won the title on a Greeves, a machine made in Essex, in 1960 and again in 1961 but in recent years the Continentals and the Japanese have taken the title.

Through the years the air-cooled, single-cylinder two-stroke has dominated the mechanical scene and in this class the European factories have battled long and hard with the Japanese.

German Honda rider Rolf Dieffenbach flies high in the 250 cc Spanish Moto Cross Grand Prix.

The first Japanese success came in 1970 when the famous Belgian Joel Robert switched from a Czech CZ to a Suzuki and retained his 250 cc title. Robert held the title in 1971 and 1972, then Sweden's Hakan Andersson won it for Yamaha. Europe hit back in 1974 when the Russian Gennady Moisseev took the crown riding an Austrian-built KTM.

Since then the championship has swayed back and forth and as we start a new decade it is still undecided, with Belgian Georges Jobe winning the 1980 championship for Suzuki, but with a European machine, a West German Maico, ridden into second place by Dutchman Wil van der Ven.

In fact the Maico that you can buy is just about the most popular 250 cc scrambler for the private rider. It is one thing to build a couple of super special works racers for your factory riders to use, it is quite another to produce similar machines for sale. And while the Japanese have been successful in providing their contracted riders with very competitive machines, they have not been quite so good at building and selling replica models.

A typical 250 cc moto crosser of the eighties is powered by the good old single-cylinder two-stroke, which produces around 40 hp at 8000 revs and drives via a six-speed gearbox.

Suspension travel, so important in soaking up the bumps, has been increased over the years until it is now around 10 inches front and rear – over double what it was in the sixties.

As smaller engines developed in efficiency and power, so the 125 cc class of moto cross grew. A European championship was started in 1973

Italian moto cross machine – the 125 cc Villa built by the Villa brothers in a small factory in Modena.

and was won by the Belgian Andre Malherbe on a West German Zundapp. Malherbe took the title again in 1974 for Zundapp and the class was so popular that it gained full World

Championship status the following year, when it was won by another Belgian, Gaston Rahier, for Suzuki.

He won again for the Japanese factory in 1976 and 1977 before giving way to the Japanese Watanabe (Suzuki) in 1978. Suzuki kept up the pressure with Harry Everts, another Belgian, who was the 1980 champion.

However, despite the World Championship tag, 125 cc moto cross has not flourished to anything like the extent that the 250 cc class has, and is virtually unknown in many countries including the United Kingdom where until 1981 the only 125 cc races were for the Senior and Expert class of schoolboy scrambling.

A sport which has been completely taken over by lightweights is trials riding and every weekend during the autumn, winter and spring up to 5000 riders compete in events all over the UK. Years ago the 500 cc and 350 cc four-strokes were the top machines for the expert trials rider.

Then, in the sixties the Spanish Bultaco factory started the trend for lightweights when they signed Ireland's trials riding wizard, Sammy Miller, to develop their 250 cc single-cylinder, two-stroke. Many doubted the wisdom of the move and said that the big, slow-revving four-stroke could not be beaten. Sammy proved them wrong and now a four-stroke in trials is as rare as an unbiased opinion in a party political broadcast.

It is true that some manufacturers have increased the capacity of their engines over the 250 cc lightweight limit (the Montesa on which Sweden's Ulf Karlson won the 1980 Trials World Championship was close to 350 cc) but trials machines remain true lightweights in spirit, with few machines tipping the scales at much over 200 lb.

As with moto cross, all are powered by single-cylinder, two-strokes tuned to give a tremendous spread of pulling power rather than top end punch. For trials are all about the ability to negotiate hazards slowly with the rider keeping his feet firmly on the footrests, and the most successful bikes are those that can go the slowest and accelerate best from near zero revs.

Typical 250 cc moto cross machine, the 1980 Suzuki RM250. Belgian Georges Jobe won the World Championship on a similar machine.

England's Chris Griffin on an Italian Fantic – one of a new breed of lightweight that are taking over in trials.

Honda, Yamaha and Suzuki have all taken an interest in trials but have never yet managed to win the World Championship which up to now has been dominated by the Spanish Bultaco and Montesa factories. However, both have been hit by the recession and falling sales in the United States and it is the Italians who are taking up the challenge for Europe

Built to balance – and for going slowly, the Beamish Suzuki RL250 trials machine.

with the Fantic and SWM factories in the forefront.

The other off-road sports with a strong lightweight element are grass-track racing and enduros. Grass-tracking, similar to speedway but run on a grass-surfaced oval instead of cinders or shale, is popular in parts of the UK and in some locations in Europe, notably France and Holland.

The 500 cc class is the major one, contested by bikes powered by speedway engines, but the 250 cc class is thriving and is the ideal stepping stone by which riders graduate to the five-hundreds. As there is no 250 cc speedway, machines are powered by a variety of moto cross engines, with the Honda a popular choice.

Currently the Italians are top dogs at international long-distance trials and this Beta 125 cc model is typical of the machines they use.

Enduros are similar to car rallies. Machines must be street legal in every way (including lights) and the events are run on a mixture of road and off-road conditions, with farm and forest tracks the favourites. The bikes are racing versions of the now popular trail machines, and with a big market in the United States the Japanese have really gone to town, with all four of the big oriental makers offering competitive enduro models in a variety of lightweight sizes.

The most famous range is the Suzuki PE models (175 and 250cc) which have established themselves as firm firm favourites, though now challenged by Yamaha, Honda and Kawasaki. For the rider who takes his sport seriously, and who has the skill to use a machine to the utmost, the Austrian KTM, Italian SWM, Canadian Can-Am and Spanish Montesa factories all offer potential winners.

The off-road motor cycle is certainly here to stay and with third world countries clamouring for cheap transport, the market for basic machines of this type seems certain to boom – and Suzuki are leading the way with what they call Farmbikes: basic, no-nonsense lightweights developed from their trail and enduro models as rugged, go-anywhere vehicles at budget prices – a sort of poor man's Landrover or Jeep.

POWER-PACKED LIGHTWEIGHTS

The growing importance of light-weight motor cycles is mirrored by the growth of the smaller capacity classes in road racing, the showpiece sport of the two-wheeled world that regularly attracts huge crowds to circuits all over the globe.

In the thirties the 250 cc class was the smallest normally held. But after the war the efficiency of lightweight engines had grown to such an extent that when the World Championship series was instigated in 1949, the 125 cc class was included along with the 250, 350 and 500 cc.

Initially disputed by the Italian Mondial, Morini and MV Agusta factories, it proved an immediate success. Even in those days these little single-cylinder, double overhead camshaft four-strokes revved to 11,000 and were capable of speeds very close to 100 mph.

Over the years the interested factories have developed faster and faster machines. The Japanese moved in to dominate the class in the sixties and early seventies and it is interesting to note that the 125 cc World Championship is the only division to be won by every one of the big four: Honda and Yamaha taking the title four times apiece, Suzuki three times and Kawasaki just once.

During the Japanese battle for supremacy some really incredible lightweights were designed and raced. Honda first won the title in 1961 with a four-stroke twin. But this was easily outpaced in 1963 by the rotary valve Suzuki two-stroke twin – developed by Ernst Degner who joined them after defecting from East Germany where he worked and raced for MZ.

To counter the growing two-stroke threat, Honda then produced a four-cylinder 125 cc racer (in 1964) and, a year later, a five-cylinder model which had pistons the size of cotton reels, revved to an ear shattering 20,000 and produced 35 hp.

Produced at enormous cost, it was good enough to stem the two-stroke tide for just one year when Luigi Taveri regained the 125 cc crown for Honda in 1966. Then Yamaha took over, first with a two-stroke twin and then with an

Winner of five World Championships in three years, Kork Ballington in action on his 250 cc Kawasaki in the Venezuelan Grand Prix.

The tiny size of 50cc racing machines is well shown in this shot taken soon after the start of the Dutch TT.

incredible little water-cooled, four-cylinder, two-stroke on which, in 1968, Bill Ivy lapped the Isle of Man TT circuit at over 100 mph – a record that is likely to stand forever, because the class was discontinued at the TT after 1974.

The Japanese were losing interest in the class which was won in 1969 by Dave Simmonds on a Kawasaki which he prepared himself, and after a win by West German Dieter Braun on an ex-works Suzuki in 1970 the Europeans regained control, with successes in 1971 and 1972 by the little Spaniard, Angel Nieto, riding a Barcelona-built Derbi two-stroke twin.

When Derbi pulled out through lack of cash Sweden's Kent Andersson took over on Yamaha twins to take the title in 1973 and 1974, but the Italian factories saw an opening and the following year a beautiful little disc valve, two-stroke twin designed by West German engineer, Jorg Moller, built by Morbidelli and ridden by Paolo Pileri, put Europe well and truly back in the 125cc picture.

Since then Moller-inspired machines built by Morbidelli, MBA and Minarelli have dominated the class – the engines now giving over 40 hp; enough to propel the bikes at close to 140 mph under favourable conditions.

While all this activity in the 125cc class was going on, 50cc racing was born – and elevated to World Champion status in 1962 when the title

56

was won by Ernst Degner on a Suzuki which he largely designed himself.

Although technically interesting, the tiddler class has not proved tremendously popular with spectators or manufacturers. The Japanese showed an early interest, with Honda coming in to challenge Suzuki, and both were developing three-cylinder models in the late sixties when they decided it simply was not worth the effort, and pulled out.

From a technical standpoint the heyday of 50 cc racing was in fact the

sixties. In those days anything went and Honda raced an overhead camshaft twin with a ten-speed gearbox! For these little engines produced their maximum power over such a narrow rev-range that it was essential to have ultra-close ratio gearboxes, so that the rider could change up or down the split second that the rev-counter needle began to waver out of the power band.

The factory 50 cc Suzukis had even more gears – their riders had a choice of 12 and 14 speed clusters which

Frenchman Guy Bertin (Motobecane) leads Spaniard Angel Nieto (Minarelli) during one of the many good dices they had during 1980.

were fitted according to the type of circuit and the personal choice of the rider. The engine was a water-cooled, two-stroke and speeds topped the 100 mph mark; works rider Yoshimi Katayama actually lapping the fast but hilly Belgian Grand Prix circuit at Francorchamps at fractionally over the magic ton in 1967.

A mechanic works on a 50 cc van Veen Kreidler. The single-cylinder is water-cooled. Top speed is about 120 mph.

Not everyone knows it but the great Barry Sheene raced a fifty back in 1971. Here he is seen in action on a van Veen Kreidler in the Spanish Grand Prix.

These incredibly complex little machines cost a fortune to build and race. The West German Kreidler factory, who had pioneered the class in Europe, simply could not afford to build machines to rival the Japanese, and pulled out. Eventually Honda and Suzuki realized they were spending more money on the class than they could afford and they too withdrew from 50 cc racing.

At this time the governing body of the sport, the Federation Internationale Motocycliste (FIM) decided to bring in rules designed to keep the cost of racing down and to channel development along lines that could benefit the normal road-going motor cyclist.

To achieve this the FIM decreed that the 50 cc class would be limited to single-cylinder machines – and that the 125 cc and 250 cc bikes would be allowed a maximum of two cylinders and that all three divisions would be restricted to six-speed gearboxes.

With the Japanese out and with these new regulations as guidelines, Kreidler, with sponsorship and help from their Dutch importers van Veen who actually ran the team, returned to 50 cc racing. With challenges from the Spanish Derbi and Bultaco teams, they have been the mainstay of the class ever since.

It says a lot for the painstaking development and thought that goes into these simple little machines that they are now far faster and easier to ride than the mechanical marvels, produced virtually regardless of expense, by the Japanese in the sixties.

The 1980 Kreidler weighed just 120 lb, stood knee-high, and with an engine developing 22 hp had a top speed of around the 120 mph mark under favourable conditions – and a strong wind makes a lot of difference to a fifty!

Spain's Ricardo Tormo switched to the van Veen Kreidler team for 1980 but could not regain the title he won in 1978. Here he is in action in the Dutch TT.

But while the smaller machines may fascinate, the mainstay of the lightweight class in racing has always been the 250 cc and the battles fought for honours in this category have been as tough and technically interesting as any in motor cycle sport.

The Italian Benelli factory had a four-cylinder racer as early as 1940 and Honda first made their name in motor cycle racing with a similar model in the early sixties. As competition increased they countered with a six-cylinder 250 cc racer. This proved to be one of the most successful racers of all time and was ridden to many victories by Mike Hailwood.

In fact it was Hailwood's brilliant riding which kept the faster four-cylinder Yamahas at bay in 1967 when the two-strokes were getting the upper-hand. Following the Japanese withdrawal and the FIM's limit on cylinders and gears the class spent a brief period in the doldrums before Yamaha revitalized it with their brilliantly simple, yet incredibly fast, two-stroke twins, the machines that have formed the backbone of the class for over a decade.

The first models were produced for racing in the United States and were developed from a successful Yamaha sports machine. Initially they were air-cooled and Rod Gould, now in partnership with Mike Hailwood in a motor cycle business in Birmingham, was the first man to win the World Championship on an 'over the counter' Yamaha racer.

For the first time in the history of the class a private rider was able to buy a machine capable of competing with factory machines. It was a milestone in racing and it is to their credit that every year since then Yamaha have built and sold a batch of racers which can, in the right hands, win Grands Prix.

No tubular frame for this lightweight
- the 125 cc **MBA** twin has a welded
box section monococque frame that
doubles as a fuel tank.

Unusual layout of the factory
Kawasakis that dominated the 250 cc
class in recent years is shown in this
shot. The water-cooled cylinders are
placed one behind the other tandem
style. Note too the rear suspension
with a single unit mounted vertically
behind the gearbox.

West German Toni Mang, 1980
250 cc World Champion, heads for
victory in the British Grand Prix.

The Italian Aermacchi factory, bought by Harley Davidson and later raced under the American name though built in Italy, were the first to respond – with what was in fact a straight copy of the Yamaha plus some improvements.

Riding these machines, Walter Villa of Italy won the title three years in succession (1974, 1975 and 1976). Then Morbidelli fielded a Jorg Moller inspired twin, a scaled-up version of their all-conquering 125 cc, and Mario Lega took the title on this machine in 1977.

Kawasaki had been playing about with an interesting two-stroke twin for a number of years without too much success. In their design the two cylinders are placed one behind the other, tandem style, to reduce frontal area, with the crankshafts geared together. In 1978 they decided to have a real crack at the championships and signed on South African Kork Ballington and Australian Gregg Hansford. Ever since, they have dominated the class, with Ballington winning the title the first two years, and West German Toni Mang taking over

when Ballington missed several races in 1980 due to illness.

Kawasaki now produce a limited batch of these machines each year which go to selected importers around the world. The engines produce over 60 hp, rev to 12,000, drive via a six-speed gearbox and propel machine and rider at speeds over 150 mph. The lightweight class has come a long way since the first simple competition models of the twenties!

INDEX

Aermacchi 63
Albion 12
Alpine 22
Andersson 49, 56

Ballington 55, 63
Benelli 23, 24, 30, 31, 60
Beta 26, 54
Betzelbacher 48
Bickers 48
Brampton 12
Braun 56
Bristol 9
BSA 8, 10, 13, 23
Bultaco 50, 51, 59
Burman 12
Butterfield 12

Cagiva 37
Can Am 54
Canning Town 7
Chater-Lea 12
Clark 9
Ciaco 20
Cologne 40
Connaught 12
Coventry 7
Cox 9
Cross 12
CZ 49

Davison 7
Degner 55, 57
Derbi 56, 59
Dieffenbach 48
DKW 23
Douglas 9
Dream 35
Druid 12
Dutch TT 56

Eagle 23
Egerton 9

Everts 1, 50
Excelsior 7, 13
Express 19, 20

Fantic 44, 51
FIM 59, 60
Francorchamps 59, 60

Gilera 23, 29
Gosport 12
Gould 60
Graduate 22
Greaves 48
Griffin 51

Haden 12
Hagon 44
Hailwood 60
Hansford 63
Harley-Davidson 63
Herne Hill 7
Honda 1, 16, 19, 20, 22, 23,
    24, 35, 36, 43, 44, 51, 54,
    55, 57, 59, 60

Isle of Man TT 7, 56
Ital-jet 38, 44
Ivy 56

James 9, 11
Jawa 16, 17, 36
Jobe 49
John O'Groats 9
Junior 12

Karlson 50
Katayama 57
Kawasaki 14, 18, 22, 35, 43,
    54, 55, 56, 62, 63
Kreidler 58, 59, 60
KTM 49, 54

Lambretta 19
Land's End 9
Lega 63
Levis 12
Liberty 12
Los Angeles 46

Magnum 38, 39
Maico 48, 49
Malaguti 38, 40, 41
Malherbe 49
Mang 634
Maplestone 12
Matchless 7

Maxi 21
MBA 56
Miller 50
Minarelli 56
Minerva 7
MMC 7
Mobylette 14
Moisseev 49
Moller 56, 63
Mombasa 6
Mondial 55
Montesa 50, 51, 54
Morbidelli 56, 63
Morini 37, 55
Motobecane 14
MV Augusta 55
MZ 16, 22, 23, 36, 37, 55
Mustang 17

New Hudson 12
Newey 12
New Imperial 8, 13
Nieto 56
Norton 12
Noyce 44

Passola 21
Pileri 56
Puch 21, 38, 39

Rahier 50
Robert 49
Rond 50
Ronchino 40, 41
Royal Enfield 12, 13

Scorpion 35
Sheene 58
Singer 12
Sturmey-Archer 12
Suzuki 15, 16, 20, 22, 23, 27,
    28, 32, 33, 34, 35, 40, 42,
    43, 49,, 50, 51, 54, 55, 56,
    57, 59
SWM 51, 54

Taveri 55
Technomoto 38
Tormo 60
Tracker 23
Triumph 7, 12, 13
Twiggie 19

Union 12
Uni-Trak 43

Van der Ven 49
Van Veen 58, 59
Vespa 19, 20, 21
Villa 49, 63
Villiers 12

Watanabe 50
Werner 9

Yamaha 16, 17, 21, 22, 23, 24,
    25, 30, 32, 33, 35, 40, 42,
    43, 44, 49, 50, 51, 54, 55,
    56, 60, 63

Zundapp 12, 37, 49